Hilarious
Harry has

1

huge horn
for hearing

Tubby takes

2

tremendously
tasty tacos

Sammy sings silly songs to

3

slithering snakes

Burt bounces over

4

bursting buckets
of blueberries

Jerry juggles

5

jugs of juniper berry juice

Penny piles

6

putrid pickle pies

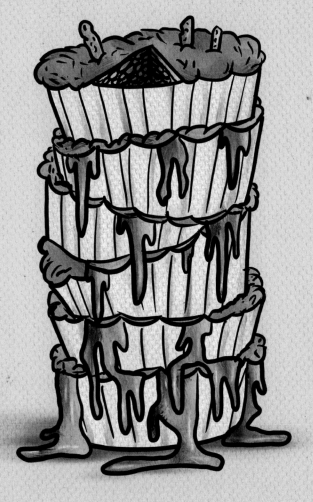

Little Lenny licks

7

lemons for luck

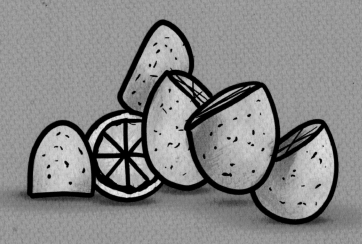

Finn
finds

8

flying
fish
funny

Glumly grumbles as he guards

9

gorgeous gourds

Otto observes an odor from

10

overripe onions

Clumsy Cole collects 11 ice cream cones

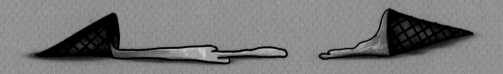

Stickers snickers at

12

spiders sipping soda
from straws

Abby acquires
13
acorns from an angry squirrel

Mustachio moves
14
mugs of milk with
his mustache

Desmond dines on 16 delectably delicious desserts

Charming Chelsie consumes 17 chilled cherries

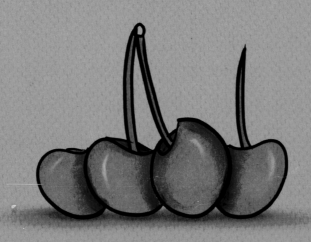

Kenny keeps

18

kazoos from a
kind kangaroo

Big Billy
boldly
bounces
19
bouncy balls

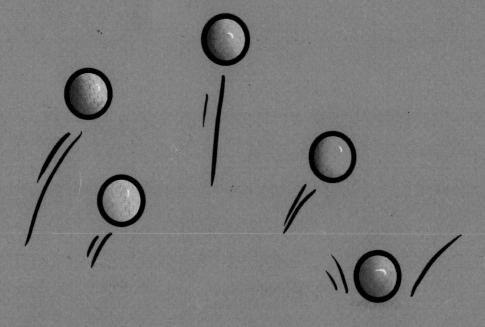

Trent and Teddy tried on

20

tremendously terrific ties today

DZOBEL
ILLUSTRATION

Special thanks to my family: Shaina, you always support all of my ideas, no matter how crazy it is. Landon, you are always my creative inspiration. Lila, you are my measure for whether an idea is good or not - if you like it, then it's good. Mom, thank you for pushing me to write and create these books. Without your everlasting support this book would not have happened. Dad, thanks for logic and laughter.

David Zobel
Designer/Illustrator

dzobel.com
david@dzobel.com

443.691.3914

Made in the USA
Columbia, SC
06 June 2017